Percentage Simple

A Short Tutorial on How to
Work Out Percentages

Anthony S. Georgilas

Percentages Made Simple

A Short Tutorial on How to Work Out Percentages

Anthony S. Georgilas

To my late father.

Contents

Preface

I decided to write this book because I have realised that many of us have some trouble understanding what percentages are and how to work them out.

For instance, a friend of mine has called me a few times asking, "if something costs 45 euros after a 20% discount, how much did it cost before the discount?" or "if I buy something for 25 euros, how much VAT do I pay?" etc.

Not only adults have difficulty calculating percentages, but also young children who have been taught them at school relatively recently.

I'll try to make things as straightforward as possible in this small book. After all, working out percentages isn't rocket science!

Introduction

Every day we see headlines like "50% off sale this week" or "the unemployment rate in Greece has risen to 21%" or "our income has dropped 35%, because of the austerity measures, since 2010" or "there's 15% chance of rain today" etc.

Percentages play an essential part in our lives, and learning how to operate them is vital.

But, where did the term "**per cent**" come from? According to Wikipedia[1]:

> In Ancient Rome, long before the existence of the decimal system, computations were often made in fractions which were multiples of 1/100. For example, Emperor Augustus levied a tax of 1/100 on goods sold at auction known as *centesima rerum venalium*. Computation with these fractions were similar to computing percentages. As

---

[1] http://en.wikipedia.org/wiki/Percent

denominations of money grew in the Middle Ages, computations with a denominator of 100 become more standard and from the late 15th century to the early 16th century it became common for arithmetic texts to include such computations. Many of these texts applied these methods to profit and loss, interest rates, and the *Rule of Three*. By the 17th century, it was standard to quote interest rates in hundredths.

The symbol '%' reads **"per cent"**, which is short for the Latin phrase **"per centum"** and means *"in every hundred"*. The words "century", "centennial", "entigrade", "centimetre" etc. have the same stem *centi-*.

Simply speaking, a percentage is a fraction that has a denominator of 100. So, for example:

$$23\% = \frac{23}{100}$$

In general:

$$x\% = \frac{x}{100}$$

Well, let's go on!

Converting Decimals and Fractions into Percentages

To convert a decimal into a percentage, we move the decimal point two places to the right and append the '%' sign. If the number has fewer than two decimal places, we add zeros.

Examples:

$$0.35 = 35\%$$

$$1.2 = 120\%$$

$$0.175 = 17.5\%$$

Easy, right?

To convert a fraction to a percentage, we either convert the fraction to its equivalent with denominator 100, multiplying or dividing its terms with the same number, or we simply divide its terms and convert the answer into a percentage, as shown above. We can also multiply the fraction with 100%, which is equal to $\frac{100}{100}$ or 1.

Examples:

$$\frac{3}{5} = \frac{3 \times 20}{5 \times 20} = \frac{60}{100} = 60\%$$

$$\frac{5}{8} = 0.625 = 62.5\%$$

$$3 = \frac{3}{1} = \frac{3 \times 100}{1 \times 100} = \frac{300}{100} = 300\%$$

$$6 = 6 \times 1 = 6 \times 100\% = 600\%$$

Pretty simple, huh?

Converting Percentages into Decimals and Fractions

We can convert a percentage into a decimal by shifting the decimal point two places to the left, adding zeros if necessary.

Examples:

$$24\% = 024.\% = 0.24$$

$$125\% = 125.\% = 1.25$$

If we want to convert a percentage into a fraction, we drop the '%' sign and write the percentage as a fraction with denominator 100. Then, if possible, we express the fraction to its lowest terms.

Examples:

$$45\% = \frac{45}{100} = \frac{45 \div 5}{100 \div 5} = \frac{9}{20}$$

$$17.5\% = \frac{17.5}{100} = \frac{17.5 \times 10}{100 \times 10} = \frac{175}{1000} = \frac{175 \div 25}{1000 \div 25} = \frac{7}{40}$$

Not very complicated, once you get the hang of it.

Comparing Quantities

To compare quantities, we must first convert both of them into percentages.

Finding What Per Cent of a Quantity is Another Quantity

To express a number as a percentage of another one, we write the first as a fraction of the second and then convert the fraction into a percentage as shown in How to Convert Decimals and Fractions into Percentages.

Example 1: Find what percentage of a year are 3 months.

$$\frac{3 \text{ months}}{12 \text{ months}} = 0.25 = 25\%$$

Warning: The two numbers must be expressed in the same unit.

Example 2: In a class of 25 students, there are 15 girls and 10 boys. Find the percentages of girls and boys in the whole class.

For the girls:

$$\frac{15}{25} = \frac{15 \times 4}{25 \times 4} = \frac{60}{100} = 60\%$$

And for the boys:

$$\frac{10}{25} = \frac{10 \times 4}{25 \times 4} = \frac{40}{100} = 40\%$$

 Tip: We could, of course, do the subtraction 100% - 60% = 40%.

How to Compare Quantities

Now, it's easy to compare two or more quantities. We simply convert them into percentages.

Example 3: Mary and Melina earn $1100 and $900 per month, respectively. Last month, Mary saved $55 and Melina $65. Who saved a bigger percentage of her salary?

Mary saved:

$$\frac{55}{1100} = \frac{55 \div 11}{1100 \div 11} = \frac{5}{100} = 5\%$$

Melina saved:

$$\frac{65}{900} \approx 0.0722 = 0.0722 \times 100\% = 7.22\%$$

We see that Melina, in spite of having a smaller salary, saved a bigger percentage of it.

Finding a Percentage of a Quantity

To find a percentage of a quantity, we convert the percentage into a fraction or decimal (see Converting Percentages into Decimals and Fractions) and then multiply by the quantity.

Example 1: Find how much VAT you must pay for a computer that costs $450, if the VAT rate is 23%.

$$23\% \times 450 = 0.23 \times 450 = 103.50$$

So, you must pay an extra $103.50 VAT. In total, you'll pay $450.00 + $103.50 = $553.50 for the computer.

Example 2: If there are 120 passengers aboard an aeroplane and 15% of them are children, determine the number of children on board the plane.

$$15\% \times 120 = 0.15 \times 120 = 18$$

So, there are 18 children aboard the aeroplane.

Not very complicated, is it? Let's do another one:

Example 3: A small town has 1870 female and 1530 male residents. If 80% of the females and 70% of the males have a driver's licence, find the percentage of the residents who can drive a car.

The total number of residents is:

$$1870 + 1530 = 3400$$

The number of females who can drive a car is:

$$80\% \times 1870 = 0.8 \times 1870 = 1496$$

The number of males who can drive a car is:

$$70\% \times 1530 = 0.7 \times 1530 = 1071$$

The total number of residents who have a driver's licence is:

$$1496 + 1071 = 2567$$

So, the percentage of residents with a driver's licence is:

$$\frac{2567}{3400} = 0.755 = 75.5\%$$

(See Comparing Quantities).

Finding a Quantity When We Know a Percentage of It

To find a quantity when we know a percentage of it, we convert the percentage into a fraction or decimal (see Converting Percentages into Decimals and Fractions) and then divide the quantity by the fraction or decimal.

Example 1: Find 100% of a sum of money if 40% of it is $120.

$$\frac{120}{40\%} = \frac{120}{0.4} = \frac{120 \times 10}{0.4 \times 10} = \frac{1200}{4} = 300$$

or, using fractions:

$$\frac{120}{40\%} = \frac{120}{\frac{40}{100}} = 120 \times \frac{100}{40} = \frac{12000}{40} = \frac{12000 \div 40}{40 \div 40} = 300$$

Example 2: Find 20% of a liquid if 8% of it is 150 mL.

First off, we'll find 100% of the liquid:

$$\frac{150}{8\%} = \frac{150}{0.08} = \frac{150 \times 100}{0.08 \times 100} = \frac{1500}{8} = 1875$$

Then, we'll find 20% of the liquid (see Finding a Percentage of a Quantity):

$$20\% \times 1875 = 0.2 \times 1875 = 375$$

So, 20% of the liquid is 375 mL.

Example 3: There was a 30% discount in a sale. Siobhán paid €28 for a pair of jeans in the sale. What was the original price of the jeans?

Using the Unitary Method[2]

$$28 \text{ euros represents } 70\%$$

$$28 \div 70 = 0.40 \text{ euros represents } 1\%$$

$$0.40 \times 100 = 40 \text{ euros represents } 100\%$$

So, Siobhán bought the jeans for €40.

---

[2]https://en.wikipedia.org/wiki/Unitary_method

Percentage Change

In this chapter, we'll deal with questions of the kind "if someone with wages of $850 per month gets a 3% raise, what will his new salary be?" or "how much will a pair of shoes marked at £40 cost after a 30% discount?" or "if a car is purchased for €15,900 and sold for €12,300 after three years, what's the owner's loss?".

Increasing and Decreasing Quantities

To increase a quantity by a percentage, we convert the percentage into a decimal (see Converting Percentages into Decimals and Fractions), add it to 1 and then multiply the result by the quantity.

We'll make things clear with an example:

Example 1: Increase €720 by 8%.

If we increase an amount by 8%, we'll have $100\% + 8\% = 108\%$ of the amount. So, we multiply by $\frac{108}{100} = 1.08$ to get the increased amount. Thus:

$$720 \times 1.08 = 777.60$$

To decrease a quantity by a percentage, we convert the percentage into a decimal (see Converting Percentages into Decimals and Fractions), subtract it from 1 and then multiply the result by the quantity.

Example 2: Decrease €12,500 by 25%.

If we decrease an amount by 25%, we'll have $100\% - 25\% = 75\%$ of the amount. So, we multiply by $\frac{75}{100} = 0.75$ to get the decreased amount. Thus:

$$12500 \times 0.75 = 9375.00$$

That's it!

Finding a Percentage Change

In some cases, we have the original amount and the new amount, after an increase or decrease. To find the percentage change, we express the new amount as a fraction of the original amount, convert the result into a percentage (see Converting Decimals and Fractions into Percentages). Then, in case of an increase, subtract 100% from the result and in case of a decrease, subtract the result from 100%.

Let us see some more examples:

Example 3: Find the percentage increase when $250 changes to $280.

We have *original amount* = 250 and *new amount* = 280. So:

$$\frac{new\ amount}{original\ amount} = \frac{280}{250} = 1.12 = 120\%$$

Then, we subtract 100% to get 12% increase.

Example 4: Find the percentage decrease when $250 changes to $180.

We have *original amount* = 250 and *new amount* = 180. So:

$$\frac{new\ amount}{original\ amount} = \frac{180}{250} = 0.72 = 72\%$$

Then, we subtract from 100% to get 28% decrease.

Calculating the Original Quantity

To find what was the original value of a quantity before a percentage increase or decrease, we convert the result into a decimal (see Converting Percentages into Decimals and Fractions). In case of an increase, we add 1 to the result and, in case of a decrease, we subtract the result from 1. Then, we divide the quantity by the result.

Example 1: Find the original amount given that, after an increase of 35%, the price is €945.

We have:

$$35\% = 0.35$$

So, the original amount was:

$$\frac{945}{1 + 0.35} = \frac{945}{1.35} = 700$$

Example 2: Find the original amount given that, after a decrease of 7.5%, the price is €111.

We have:

$$7.5\% = 0.075$$

So, the original amount was:

$$\frac{111}{1 - 0.075} = \frac{111}{0.925} = 120$$

That's all there is to it.

About the Author

Anthony S. Georgilas was born in Greece in 1964.

In July 1988 he graduated from the University of Ioannina, School of Sciences, Department of Mathematics, Probability, Statistics and Operations Research Section[3].

He writes and speaks Greek (mother tongue) and English.

Currently, he works at the Municipality of Kalamata.

---

[3]https://math.uoi.gr/index.php/en/department/sections/probability-statistics-and-operations-research

Printed in Great Britain
by Amazon

78149829R00020